UNLEASHING
THE
WORD OF GOD

D1293509

UNLEASHING
THE
WORD OF GOD

Gene Edwards

Published by SeedSowers
Jacksonville, FL 32206
800.228.2665
SeedSowers.com

ISBN 10: 0-9797515-3-5
ISBN 13: 978-0-9797515-3-0

For 1800 years,
the chaotic arrangement
of the
New Testament
has prevented us
from understanding
the
New Testament.

WHAT IF

You speak and read fluent Greek.

You have memorized the entire New Testament... *in Greek* and can *quote* the entire New Testament...in Greek.

You will still not understand what the New Testament is saying!

Why?

Has it to do with some mystery or secret? Is Scripture *that* hard to understand?

No!

The reason we cannot understand the New Testament has to do with one very crucial thing: the chaotic order in which Paul's letters are arranged.

Yes, arrangement matters that much!

WHENCE CAME THE CHAOS?

It happened 1800 years ago. This chaotic arrangement was accepted and later became *the only way* to arrange Paul's letters.

As a result, we cannot know what the New Testament is saying. We have *never* seen the church of Century One. We do not know what happened in the first Christian century. We have no model of the first-century from which to work.

Counterwise, we have been creating Christian practices made up of short passages from the New Testament, never seeing the entire panoramic saga.

THE ENTIRE CANVAS

When will we begin to see the entire first-century in an ever forward flowing panorama? Unbroken. Uninterrupted. The drama. The whole.

What would happen to you? To me? To Christians? To the church? To future Christian history?

MEET THE MAZE

On the next page, you will see the chaotic order in which Paul's letters are arranged. *(This is the order found in your New Testament.)*

Compare this chaotic order with the proper sequence.

Why We Do Not Understand the New Testament

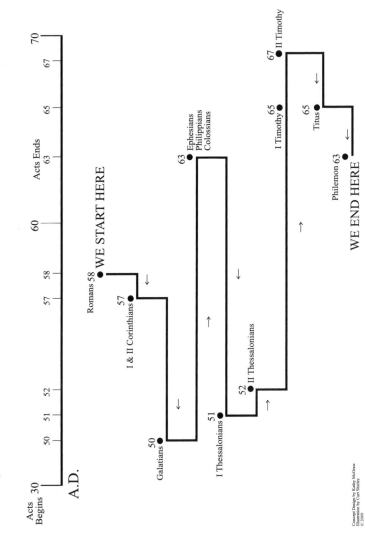

After 1800 Years
Finally, Here is the Right Way

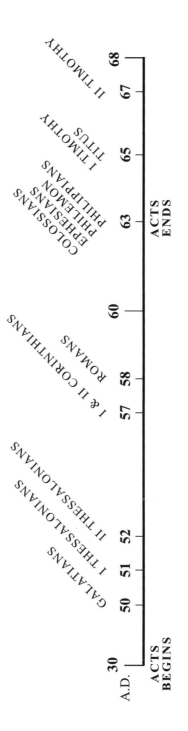

A.D. 30	50	51	52		57	58	60	63		65	67	68
ACTS BEGINS		GALATIANS	I THESSALONIANS II THESSALONIANS		I & II CORINTHIANS ROMANS			COLOSSIANS EPHESIANS PHILEMON PHILIPPIANS		I TIMOTHY TITUS		II TIMOTHY
								ACTS ENDS				

The first chart (page 4) shows how you presently read your New Testament.

Let us say that you have been reading your way through the book of Acts. You have come to the last sentence. What you read in that last chapter is that Paul is in prison, in Rome.

After Acts ends, then comes Paul's letter to the Romans. It is perfectly logical to assume that Romans continues on after the book of Acts closes.

Right?

No!

Chapter 28 of Acts and Paul's letter to the Romans have no connection! (It is Acts and Galatians which connect.) Acts is a history book and spans the years 30 through 63. Here is the disconnect: Romans (a letter) was written in 58; Acts chapter 28 ends in the year 63!

No wonder confusion reigns.

As a result of the chaotic order, the impression (of just about all of us) is that Acts and Romans *must* connect to one another. It appears that Acts flows smoothly into Romans, when actually there is no chronology here whatsoever!

You have entered the maze…be you a scholar, a theologian, a Bible teacher, or just an ordinary believer.

Romans was actually written *before* Paul was in Rome. In fact, Paul was in a country called Greece. At the time Paul wrote Romans, he had never seen Rome. Romans was written in the year 58. Paul did not arrive in Rome until the year 61.

No, Acts and Romans do not flow together.

Wait, there is more to this chaos!

After you finish reading Romans, you turn the page, and there is I Corinthians.

Surely, I Corinthians was written *after* Romans. Correct? Sanity demands it. Order demands it.

No.

Actually, even though I Corinthians is listed *after* Romans, the fact is that I Corinthians was written *before* Romans!

CORINTHIANS, BEFORE ROMANS?

Yes, I Corinthians was written in the year 57 while Paul was in Ephesus. Romans was written in the year 58, but is listed *before* I Corinthians. You have now finished reading I Corinthians. You then turn to II Corinthians. Yes,

II Corinthians was also written *before* Romans, though in your New Testament II Corinthians is listed *after* Romans.

Let us continue wandering in this maddening maze.

You finish reading II Corinthians. Then you turn to Galatians.

Surely Galatians was written *after* Romans. Surely

Galatians was written *after* I and II Corinthians. Again, sanity demands it.

No. In fact, Galatians was written *before* Romans. Galatians was also written *before* I and II Corinthians. Galatians was written in the year 50, *seven* years before Corinthians and eight years before Romans.

Pure chaos! There is no cohesiveness here. Context, dates, and places are utterly out of order and rendered useless. World history has never been dealt such a death blow as has the history of Century One.

Seeing a story, seeing anything that is orderly or sequential, is impossible. Truly, to attempt any effort of understanding Paul's letters cohesively is nothing less than a study in the impossible.

Can anyone see order in any of this? Could you keep in mind a straight, forward-moving story, or a forward-moving *anything*?

After Galatians, you turn the page. There you find Ephesians. Any of us would think those letters were written at the same time.

Actually, Galatians was written in the year 50. Ephesians was written in 63. There is a *thirteen-year gap* between those two letters: Ephesians, written thirteen years *after* Galatians!

Then you turn the page and come to Philippians, and then to Colossians. As bizarre as it might sound, Colossians was written *before* Ephesians. Further, both Colossians *and* Ephesians were written *before* Philippians. (That is, Philippians was written *after* Colossians, not *before.*) The only connection here is that all three letters were written in the year 63.

There was yet another letter written that same year. Of course, that letter is listed as the very last of Paul's letters. (Here is a clue: This is Paul's shortest letter.)

You turn the page. You discover I Thessalonians. Surely to sanity, I Thessalonians was written *after* Colossians. Much later, perhaps? After all, you are coming to the end of Paul's letters. Therefore, I Thessalonians *must* be one of Paul's later letters.

Not so!

In fact, I Thessalonians was one of Paul's earliest letters (51), written right after Galatians. I Thessalonians is Paul's *second* letter. (Perhaps you are now beginning to know what it feels like to be lost in the maze.)

Then comes II Thessalonians, written in 52! I Thessalonians and II Thessalonians were written *before* Romans, *before* I and II Corinthians, *before* Ephesians, *before* Philippians, and *before* Colossians.

Now you come to I and II Timothy.

Here is what we have passed through in the maze so far: The Roman letter was written in 58, followed by two Corinthian letters written in the year 57, then back to the Galatians letter written in the year 50, then thirteen years ahead to the Ephesian, Philippian and Colossian letters of 63, then back to 51 and 52 for I and II Thessalonians, and ending with the first letter to Timothy in 65 and II Timothy in 67!

NO MAN HAS EVER LIVED

No man has ever lived who could hold this jumbled New Testament arrangement in a straight line. Clarity is relegated to the realm of the impossible.

You have seen it for yourself. You have seen why we *cannot* understand the New Testament in its wholeness. The living *Story* disappears. Instead, verses reign. The entire first-century witness is denied. Total context is nonexistent.

And us? We are left stranded, travelers with no compass...no map!

We have accepted the fact that we cannot understand Paul's letters (because of this present chaotic order), but we do not know why. We believe we need someone with a seminary degree to teach us. What we do not realize is that seminarians are no better off than we are!

It would not matter if you knew Greek and had memorized the entire New Testament in Greek. There is no man who has ever lived who can retain this chaotic and traditional arrangement of the New Testament and at the same time see the connection and the wholeness of Paul's letters in their proper sequence.

THE CHAOS CONTINUES

You turn the page. You have left II Thessalonians and come to I and II Timothy. Since these four letters all begin with the letter "T," there must be *some* connection. Surely!

Not at all. All four of these letters have been arranged only by *length*, not by "Ts," nor by dates, nor by events. All thirteen letters are arranged by *length* alone.

But, are not I Timothy and II Timothy in order?

No. The proper order is: I Timothy, then Titus, then II Timothy.

I Timothy and Titus were written in the year 65. II Timothy was written in the year 67.

Alas, is this not finally the end of the chaos?

No, not yet. The greatest mind boggler comes last.

We come, finally, to a very short letter called Philemon.

Philemon was *not* written after II Timothy. Philemon was written in the same year Colossians, Ephesians, and Philippians were written, in 63.

Why is Philemon listed last? Once more, all thirteen letters are arranged by length. Philemon is Paul's shortest letter. Forget sanity. Length reigns.

Please look once more at the labyrinth (see page 4). Take note. This labyrinth—this maze—is what you enter every time you turn to the letters of Paul.

Should you place Paul's letters in the order Paul wrote them, the change would be, well, try it and see. Otherwise, no man could put all this jumble together and make sense of it.

That leaves us with a simple fact. None of us have understood the New Testament. The reason is simple. The solution is just as simple.

Let us suggest how simple it is. This solution will contrast proper sequence with length!

Here is the order of Paul's letters as listed in your New Testament. We will include the dating of each of Paul's letters.

Thirteen Letters, by Length

(Observe the dates and the subsequent chaos this causes.)

Letter by Length	Date Written
Romans	58
I Corinthians	57
II Corinthians	57
Galatians	50
Ephesians	63
Philippians	63
Colossians	63
I Thessalonians	51
II Thessalonians	52
I Timothy	65
II Timothy	67
Titus	65
Philemon	63

Look again at this incoherent order of dates:

58, 57, 57, 50, 63, 63, 63, 51, 52, 65, 67, 65, 63.

There is *no* hope of clarity in ever seeing what the first century looked like and what the church was like.

What happens if proper sequence steps upon the scene?

Here is the order in which Paul's letters *should* be arranged...by the order in which he wrote them. See the sensibility of the dates.

Galatians	50
I Thessalonians	51
II Thessalonians	52
I Corinthians	57
II Corinthians	57
Romans	58
Colossians	63
Ephesians	63
Philemon	63
Philippians	63
I Timothy	65
Titus	65
II Timothy	67

Note how orderly, neat, and chronological Paul's letters *could be*. Further, there is a connecting sequence of events between his letters. Out of this arrangement, *The Story* emerges as a complete whole.

50, 51, 52, 57, 57, 58, 63, 63, 63, 63, 65, 65, 67.

WHY 1800 YEARS?

Have you ever read a book that arranged each chapter by its length, rather than by forward progress of the story?

Why, after 1800 years, are we still arranging the New Testament according to length and topic chronology?

Just what is the story that is hiding in these letters, once they are in the proper order?

Right now, we are all like explorers who have never even known there was a map.

WORST OF ALL

From the first days of our conversion to Christ, you and I are told to read the Bible. Everyone does. Once. After that, few continue.

Why?

Unconsciously, perhaps, we do not enjoy reading the Bible because we cannot get hold of direction. Perhaps unknowingly, we sense this utter disarray. This, in turn, prevents us from discovering our heritage. We are unable to see those men and women—our brothers and sisters—living a life of adventure and excitement in the organic expression of church life!

So, we continue blindly living our Christian life as orphans—not knowing our family story—and we are ultimately denied our birthright.

THE NEW TESTAMENT
AND THE KORAN

There are only two books still read today
which are bound by *length*, not by sequence.
One is Mohammed's Koran.
The other is the New Testament!
They have something in common:
an ancient, heathen bookbinding practice!

FROM WHENCE CAME
THESE CHAINS?

Here is how those chains were first wrapped around the Scripture. It all had to do with one of the traditions of antiquity. Back in that era, books were rare. Books were compilations of writings on various subject. These writings were bound into a book by the length of each topic.

The famous writers of that age were men such as Horace, Cicero, Pliny, Seneca, Diogenes and, of course, Aristotle. When a bookbinder of that day bound together the writings of one of these men, the bookbinder first measured the length of each separate topic on which the author had written. The longest pieces came first, the shortest last, with all other topics being bound in descending order.

The bookbinder would treat Paul's letters exactly the same way. One day a man asked that Paul's thirteen letters be bound. The bookbinder counted the words in each of Paul's letters and discovered that Romans was the longest letter and Philemon was the shortest. Such was the birth of the present-day chaotic order of Paul's letters.

This odd arrangement continued for 1800 years and is with us to this very day. This arrangement of Paul's letters by length has become institutionalized chaos.

After all, Paul did not write philosophical treatises. Rather, Paul wrote warm, even passionate letters, letters to people with names, people caught in unusual circumstances, people in churches and crises. He wrote to some of those people by name, and some of them by the name of the church. Each letter was surrounded by circumstances and current events which provoked Paul to write. There are historical events in every letter he wrote.

Place those letters in the order
Paul wrote them and they tell
a story...one unbroken,
dramatic story.

In fact, Paul's letters, when placed together, tell *The Story*, a story which stretches from the year 50 through 67. Blend in Acts with those letters, and now *The Story* stretches uninterrupted in three dimensions from the year 30 through 70...from Pentecost until the destruction of Jerusalem. That is a revolutionary story, and it has not been told in 1800 years!

THE POWER OF THE WORD OF GOD

At last, clarity will reign. This captivity to the chaotic began in about the year 200. Let us end its 1800-year reign, *now*! In so doing, we will loose some two-thirds of the New Testament to crystal clarity and open up forty years (30–70) which are presently a blur.

Keep in mind, these are the years when the New Testament was written! The four Gospels are a story (27–30). Why not the *complete* story (27–70)? Clarity from 27 to 70 *does* unfetter God's Word.

We can only be what we ought to be when we can *see* what we ought to be.

Let us be the generation that comes to know, for the first time in 1800 years, the entire first-century story, with the blur removed. That will be a day of revolutionary insight. It will be a radical release of the divinely inspired Writ of God.

Remove the blinders. God's Word will do the rest!

On that day we will be drawing from the waters of a well previously untapped, uninterrupted, and overwhelming.

What a well to draw from!

IS CHRONOLOGY, BY ITSELF, ENOUGH?

In recent years there have come upon the scene a few chronological New Testaments. These recent editions placed Paul's letters in their proper order. Such an arrangement helps. Nonetheless, add a few additional helps and you will set off a cataclysm.

THE OVERLOOKED YEARS

Paul wrote his letters over a period of eighteen years; yet, he wrote letters in only eight of those eighteen years. That means there are *ten* years during those eighteen years when Paul *did not* write letters.

The problem is that throughout history those years when he did not write have been completely ignored. Neither in books, nor in commentaries, has any focus been given to the years in which Paul did not write letters.

THE LETTER WRITING YEARS

Please note that Paul's letters were all written between the years 50 and 67, a period of eighteen years inclusive. Nonetheless, it is important to note that there were ten of those eighteen years when Paul wrote no letters.

Here Are the Years in Which Paul Wrote Letters:

50

51

52

57

58

63

65

67

THE MISSING YEARS
(Ten out of an eighteen-year period)

Here are the missing years. Without knowing what happened in those years, clarity will elude us.

Here Are the Years in Which Paul Did Not Write:

53

54

55

56

59

60

61

62

64

66

Here Are the Years of Paul's Letters (With the Missing Years in Italics):

50, 51, 52   **57, 58**   **63   65   67**

53, 54, 55, 56   *59, 60, 61, 62   64   66*

These years are a great deal of time to leave out when seeking to understand the New Testament!!

NEEDED: MORE THAN CHRONOLOGY

When Paul's thirteen letters are all placed together, without *any* helps, you will still not see the complete canvas. There are a number of essential helps—at least a dozen—that need to be added in order for the saga to come forth.

Keep in mind, those helps are available to you *only* when you place Paul's letters in their proper order. Otherwise, those helps cannot even exist.

HOW TO RECONSTRUCT
THE FIRST CENTURY

Here are the tools needed for this "reconstruction" project. Like carpenters, bridge builders, and stone masons, you need tools. Your tools are: history, archaeology, dates, times, places, travel conditions, climate, weather, religious customs, government. All these tools come into play in "reconstructing" Century One.

Here are your construction tools:

- First, place Paul's letters in order. Add dates to each letter.
- Add the major events which took place in Rome, Israel, and the empire *between* those thirteen letters.
- Follow the names of the people as they appear. (They will appear chronologically for the very first time.) Keep up with these people: Paul, Barnabas, Titus, Timothy, Gaius, Aristarchus, Secundus, Sopater, Priscilla, Aquila, Tychicus, Trophimus.
- Give (estimated) ages to all of these people who appear.
- Use macro and micro dates, those which appear by year. Others appear by weeks and even days.
- Include travel conditions, local customs, and miles traveled. (They affect the unfolding drama.)

- Make note of what events provoked Paul to write each letter, such as events in the empire, local crises, crises in the church, events influenced outside the church that created crises in the church.

THE RESULTS?

When you have finished, you will discover that you have a clearer understanding of the New Testament, clearer than most Bible scholars have. Why? Because you have seen the first century as a complete whole. Scholars work with sections: that is, each letter as a *stand-alone* with no smooth connection to the next letter. There is a vast difference between studying each letter as a stand-alone, as over against studying the letters flowing forth into one another in sequence.

You will then be holding in your hand a three-dimensional view of the places, events, and people who passed through those forty years.

You will come to literally *see* the entire panoramic saga.

For instance:

You will see how men were trained in the first century.

You will see what gatherings in the first century were really like.

You will see how churches were planted.

In a word, you will see Century One, with its churches, its Christians. You will see it all together in one grand sweeping drama.

You will be seeing the completed canvas.

The difference may be startling, perhaps enough to start a new Reformation, perhaps a revolution!

**Can we really see *The Story*
simply by rearranging thirteen letters
and adding a few tools?**

Yes! Further, that marvelous story turns out to be *absorbing*, even an edge-of-your-seat spellbinder. It is as enthralling as any who-done-it or any adventure/drama.

The Word of God is the best of great literature. In heaven's name, let us therefore unleash the power of His Word.

There is one more mountain…perhaps the most formidable of all.

CHALLENGE YOUR OWN MIND-SET

Today we read the New Testament, and voila…we see modern-day Christianity right there. Let us deprive ourselves of that luxury. Instead, let the first century tell us *The Story*, exactly the way it happened.

In other words, break the mind-set.

A mind-set changes from generation to generation. The habit of peering into the New Testament and seeing your own day's version of Christianity is an ancient—yet generational—habit. Catholics see a pope on every page of Scripture. You and I see a pastor on every page. *The Story* lets you see neither! *The Story* shows you what really happened.

That alone will cause enough of an upheaval to keep us busy for the next two hundred years! When you and I remove the impenetrable maze, the chaos collapses and then it disappears. The modern day mind-set—the one belonging to *this* age—also dissolves.

What will finally emerge out of this?

Possibly an overwhelming revelation of Christ!

Perhaps as many as a million *verse*-based books will be outdated. Thousands of new books will be needed to replace them.

Should we dare stay with this traditional *length*-bound order of Scripture? If so, the stage is set for another 1800 years of "non-understandability."

You can prevent that disaster. How?

THESE THINGS YOU CAN DO
TO UNLEASH GOD'S WORD

Untangle the chaos. Learn the New Testament from the year 27 to the year 70. Discard the mind-set; learn *The Story*. Contrast and compare Century One with Century Twenty-One. Every letter will explode. What you find, tell it. Write it. Teach it. Explore it. Live it. All this will profoundly affect the world!

You tell *The Story*—from start to finish—*The Story* no one has seen or read or heard for 1800 years! Become its voice. Be its advocate. *Restage* those earliest days of our faith. Repopulate the first-century. Tell the world what was, what *ought* to be.

There is no law that says we must live out our entire lives with nothing but an obscure view of those "first motion" years. Give that ancient age back to us. Give it to us in one great colossal, kaleidoscopic, panoramic epic! Give it to the whole world. End the enigmatic and utterly unnecessary mystery of the case of the missing century. Demolish the maze!

Billions of dollars have been spent during the last 1800 years for the one stated purpose of understanding the Word of God. Yet, never has that Scripture, nor that vast effort, been presented with its complete context.

But you can!

THE END OF "UNKNOWABLENESS"

Remove that blur, end the "unknowableness" of the New Testament...and you will change the world!

Believe this: The years 30 to 70 and Paul's letters are as much a story as are the four Gospels. Tell both!!

You and I can walk into the first-century, see it, taste it, feel it, know it, and then tell it...*nay*, proclaim it! Be caught up in it, then unleash it! Set free the sheer power of the Word of God, which is now captive to a century which—until now—has been unknown to us.

HOW GRIPPING IS THAT STORY?

It is gripping enough to hold the attention of the lady who works at a truck stop *and* the truck driver. It is a story better than *Gone With the Wind* ever hoped to be!

Not only tell it. Yield to it! We may even see the entire schematic of the Christian way of viewing the New Testament disappear. All that, by the simple telling of a story!

Give to all of us a New Testament which allows us to see all these things. Then, having seen it and lived it, write that story! Then maybe the world can be turned upside down just one more time!

Be sure of this: The Story puts an axe to the roots of the practices we have accumulated these 1800 years of a storyless era. Paint that canvas. Paint it with words and experience.

Give us a panoramic stage, repopulated with the original characters doing what they did, not tricking our minds into seeing what we are doing today. Show us the entire landscape and seascape, and show us even the *other realmscape*. On that glad day, we will again pursue our roots.

Alert every Christian, every movement, every denomination, every Bible school and seminary: "Here is the whole. Yield to it, *or* cease declaring that you are being true to the Word of God."

The Story, the Scripture, are once more revolutionary. Yes, *that* powerful!

Shake the world with it!

Best of all, dear Christian, this is a do-it-yourself project. *You* can discover all this for yourself. Go build your model of that century which gave us the New Covenant of God's Word.

It goes even beyond that.

The Story is so powerful, yet so new to us, once you have grasped it, it will grasp *you*.

Once it has grasped you…

"You need no man to teach you!"

You! Go shake the world!

39

NEXT STEP?

After reading *Unleashing the Word of God*,

you will want to read its companion,

Revolutionary Bible Study

(A Revolution in Bible Study for a New Generation).

INTERESTED?

THERE ARE OTHERS.

CONSIDER:

SHALL WE WAIT
ANOTHER 1800 YEARS?

What if men would have had a first-century model back during the formative years around 330–500? If such a model had existed, the present-day practice of Christianity would never have existed. What if today every seminary student had to create his own first-century model, a model of the early church, a model beginning at Pentecost and going on to Patmos?

Today, there *is* a model.

WELCOME TO
REVOLUTIONARY BIBLE STUDY!

This book is accompanied by a DVD that is intended to help you in this grand new venture of unleashing the Word of God.

The companion book takes you through every year from 30 through the year 70 in an uninterrupted, forward-flowing saga. These are the years which changed the world and birthed the New Testament. It is *all* there.

A CALL FOR A BAND OF BRAVE MEN AND WOMEN

Have you ever heard of a group of magnificent people called the Lollards? Back in the late 1300s, there were no printing presses. All books were handwritten. There was no Bible in English anywhere in England. All were in Latin. John Wycliffe made possible a new handwritten New Testament, *in English*. (There had never been such a thing before.)

A group of men—later called Lollards—took these few handwritten New Testaments and, traveling throughout England, read aloud to the masses and taught them. This was revolutionary! To hear the New Testament in their own language! Finally, the common people could hear the New Testament in a language they could understand.

The result? The Lollards are the ones who ignited the fuse for the Reformation.

Another such group of believers is needed today: those who, for the first time, will make that same Book clearer than ever before to God's people. Such a people will be armed with a New Testament that is crystal clear; they will be armed with *The Story*; they will be armed with a model. They will be a new generation with the burden, dedication, and fervor of the original Lollards.

Lollards Again, Anyone?

SeedSowers
800-228-2665 (fax) 866-252-5504
www.seedsowers.com

Revolutionary books on Church life

Beyond Radical *(Edwards)* 8.95
How to Meet In Homes *(Edwards)* 11.95
An Open Letter to House Church Leaders *(Edwards)* 5.00
Revolution, The Story of the Early Church *(Edwards)*..................12.95
The Silas Diary *(Edwards)*...10.95
The Titus Diary *(Edwards)*...10.95
The Timothy Diary *(Edwards)* 10.95
The Priscilla Diary *(Edwards)*...10.95
The Gaius Diary *(Edwards)* 10.95
The Organic Church vs. The New Testament Church *(Edwards)*......7.50
Problems and Solutions in a House Church *(Edwards)* 7.50
Why So Many House Churches Fail and What to Do About It *(Edwards)*   5.00

An introduction to the deeper christian life

Living by the Highest Life *(Edwards)* 10.95
The Secret to the Christian Life *(Edwards)* 10.95
The Inward Journey *(Edwards)*..12.95

Classics on the deeper christian life

Experiencing the Depths of Jesus Christ *(Guyon)*..........................9.95
Practicing His Presence *(Lawrence/Laubach)*.................................9.95
The Spiritual Guide *(Molinos)* 9.95
Union With God *(Guyon)* 9.95
The Seeking Heart *(Fenelon)*...10.95
Intimacy with Christ *(Guyon)*..10.95
Spiritual Torrents *(Guyon)*..10.95
The Ultimate Intention *(Fromke)*...10.00
One Hundred Days in the Secret Place *(Edwards)* 13.99

In a class by Themselves

The Divine Romance *(Edwards)*...12.99
Christ Before Creation..9.95

New Testament

The Story of My Life as Told by Jesus Christ *(Four gospels blended)* 14.95
The Day I was Crucified as Told by Jesus the Christ....................14.99
Acts in First Person *(Book of Acts)* 9.95

Commentaries by Jeanne Guyon

Genesis Commentary 10.95
Exodus Commentary 10.95
Leviticus - Numbers - Deuteronomy Commentaries 12.95
Judges Commentary 7.95
Job Commentary 10.95
Song of Songs *(Song of Solomon Commentary)* 9.95
Jeremiah Commentary...7.95

(Prices subject to change)

The Chronicles of Heaven *(Edwards)*

The Beginning	9.95
The Escape	9.95
The Birth	9.95
The Triumph	9.95
The Return	9.95

The Collected Works of T. Austin-Sparks

The Centrality of Jesus Christ	19.95
The House of God	29.95
Ministry	29.95
Service	19.95
Spiritual Foundations	29.95
The Things of the Spirit	10.95
Prayer	14.95
The On-High Calling	10.95
Rivers of Living Water	8.95
The Power of His Resurrection	8.95

Comfort and Healing

A Tale of Three Kings *(Edwards)*	9.99
The Prisoner in the Third Cell *(Edwards)*	9.99
Letters to a Devastated Christian *(Edwards)*	7.95
Exquisite Agony *(Edwards)*	10.95
Dear Lillian *(Edwards) paperback*	5.95
Dear Lillian *(Edwards) hardcover*	9.99

Other Books on Church Life

Climb the Highest Mountain *(Edwards)*	12.95
The Torch of the Testimony *(Kennedy)*	14.95
The Passing of the Torch *(Chen)*	9.95
Going to Church in the First Century *(Banks)*	6.95
When the Church was Young *(Loosley)*	8.95
Church Unity *(Litzman,Nee,Edwards)*	10.95
Let's Return to Christian Unity *(Kurosaki)*	10.95

Christian Living

The Christian Woman . . . Set Free *(Edwards)*	12.95
Your Lord Is a Blue Collar Worker *(Edwards)*	8.95
The Autobiography of Jeanne Guyon	19.95
Final Steps in Christian Maturity *(Guyon)*	12.95
Turkeys and Eagles *(Lord)*	9.95
The Life of Jeanne Guyon *(T.C. Upham)*	17.95
Life's Ultimate Privilege *(Fromke)*	10.00
All and Only *(Kilpatrick)*	8.95
Adoration *(Kilpatrick)*	9.95
Release of the Spirit *(Nee)*	6.99
Bone of His Bone *(Huegel) modernized*	9.95
You Can Witness with Confidence *(Rinker)*	10.95